Deep Learning Manual:

Foundations

First Edition

Paul Ste

Part of a series including:

Deep Learning Manual: the knowledge explorer's guide to self-discovery in education, work, and life, by Arthur J Murray (2016)

Copyright © 2017 by Paul Stephen Prueitt
All rights reserved. No part of this publication may be reproduced, distributed, or transmitted in any form or by any means, or stored in a database or retrieval system, without the prior written permission of the copyright owner.

ISBN-10: 0692798021
ISBN-13: 978-0692798027

blished by: Applied Knowledge Sciences Press
ted in the United States of America

Dedicated with love to

Mary Lynn White

Contents

Foreword

It's always exciting to see a movement take shape and transform the world in some way. And it's especially exciting when several different movements come together to create an even greater transformation. It's from this rare type of convergence that a new, old way of learning has emerged. "New old" in that it takes a fresh look at how most learning occurred prior to the Industrial Age. We're talking about the re-discovery of deep learning, which is learning at a deep structure, rather than at a surface, level.

This transformation was brought about by two separate but related movements. The first was the increasingly popular use of deep learning approaches leading up to the current advances in machine learning, a key branch of the field of artificial intelligence (AI) .

The second was a grass roots movement arising out of pent-up frustration with how basic school subjects, especially mathematics, are taught. The focus on rote memorization and obsession with numerical grades (score a 70.0 and you pass, 69.9 and you fail) has in many cases created what have come to be known as "acquired learning disabilities." In such cases the learning process is inhibited, or even worse, completely shut down as a result of the buildup of negative classroom experiences over time. The long-term result is often fear of, or disdain for, a particular topic. The good news is that in seeking a solution, long-forgotten learning approaches have been re-discovered and are being re-applied, with very promising results.

This unique set of circumstances has led to the publication of this manual.

For over thirty years, Dr. Paul Stephen Prueitt has been at the forefront of these trends, both as a research scholar and educator, and has played an active role in their convergence. The transformations he has achieved in working with students in the classroom, particularly with incoming college freshmen not majoring in science, engineering or mathematics, are nothing short of astonishing.

This manual was initially designed for use in a "Second School" learning environment, in which learning is a combination of guided mentoring, self-discovery, and peer-oriented discussion in small groups aided by social media. However, readers not registered in a Second School class can still use this manual to start experiencing this new style of learning.

The chapters are short, focusing on one foundational principle of mathematics at a time. You'll find that most of your time using this manual will be spent thinking, with very little time reading – the complete opposite of how most classes are taught.

But as you'll see, the "thinking" periods are also short, lasting about ten minutes. Think. Write. Reflect on what you've written. Take a break and do something totally different. Then come back to the topic, looking at it from a slightly different perspective. Rinse and repeat.

Deep learning is a gift that just doesn't last a lifetime. The more you use it, the more you'll realize that there's no limit to what you can discover. Best of all, it's not like cramming your head full of facts and dates and names and places. It's real, experiential knowledge you can feel deep in your bones – in your very being.

Give it a try and experience it for yourself. You won't be disappointed. In fact, you may never look at learning in the same old way again.

- Dr. Art Murray
  Editor and Publisher
  AKSciences Press

Part One: Foundations for Non-Stem Mathematics

"Foundations" **open to you a new way to conceptualize your math class.**

Everyone enrolled in math class could benefit from deep learning methods. This manual provides some of what is missing in our common perception about math, and the math class.

This manual is written for those who deserve a second chance in math class. For those who already see into the foundations of mathematics, the book is something extra.

Deep learning methods enhance natural curiosity as well produce increases in motivation.

Listing topics and separating that list into two creates a beginning.

A sense of location is found.

Take charge. Open your perception.

Anyone may join the second school movement. But membership in the second school is governed by rules. Various social manifestations will occur, and define how you interact with *social learning media.*

Misbehavior, for example, results in loss of membership.

Polls are required to advance within the school. These polls are not testing your skill at finding the right answer. Rather, they identify where you are located in a topic map. The topic map is constructed algorithmically as a model about your possible perceptions of mathematics.

You will learn things that you are not expecting to learn. Profiles in individual learning are evolved, and when membership is lost, the profiles are no longer within the system. Your identity is no longer within the system.

Data analytics is used to understand human leaning behavior and to construct profiles based on proper measurement.

Goals

Goal 1: Learn how to write college level mathematics, as class notes and as private inquiries, using what we call "topic trajectories."

Goal 2: Strengthen arithmetic skills by re-developing arithmetic in a base other than base 10.

Goal 3: Learn the full and proper notation necessary to talk about various sets, such as domain and range, solution sets and replacement sets.

Goal 4: Become self-directed in the study of any freshman college course.

Goal 5: Internalize topics, such as replacement sets and solution sets.

Goal 6: Provide introductions to polynomial, exponential, and logarithmic functions and functional composition.

First Two goals

Learning to write mathematical concepts is a difficult task, for many reasons. The difficulties flow from basic human nature.

Conjecture: When one's experiences in math class are not positive, human nature creates a resistance to learning.

For many of us, math class is just not a positive experience.

Our society is at an impasse. Everyone understands our education systems are not working for most of us. Second school philosophy makes learning to write as simple as possible. With this skill you are able to change how you perceive topics in the curriculum.

One way to learn how to write is to write about an interesting beginning curriculum involving positional notation and number bases. This is the first lesson, as presented in the next pages.

Rediscovery of arithmetic and set theory

Consider the integer 1304 in base 10.

$$(1304)_{10} = 1*10^3 + 3*10^2 + 0*10^1 + 4\times 10^0$$

Note that $10^0 = 1$.

Yes the exponent is 0 and the "*" does stand for multiplication. Learning to see into handwritten class notes is essential.

Handwritten messages are often more difficult to read, but additional meaning is always communicated. The trick is to see what cannot be expressed in a typed form

In these pages, you will be guided in developing a new skill, which you might use to send handwritten messages to a Circle of Mentors.

You should expect your college teacher to use a chalkboard and write in ways that you sometimes barely understand.[1] Notetaking and handwritten expressions, called topic trajectories, help in the two-way communication that is required for best learning experiences.

---

[1] Those of you in the room that agree with this statement, please raise your hand.

Writing to Learn

Taking class notes is a good thing. But who really does this? About 20%. You will learn to write clearly about topics and post these notes into a dedicated learning platform.

Your posts are examined by algorithms and second school mentors. The result is that your post becomes part of a larger social "game". The game is not like anything you now know about. Only those who participate know.

Looking at handwriting is different than watching a video or looking at pages in a textbook. Producing handwritten notes requires internalization of concepts.

We are asking you to write like a professor. This does not make you a master of the material, but your ability to write increases your understanding about your location within this game.

When test time comes you are able to explain the pathway you took in producing your answer. This will not resolve all of the issues we all have with math class, but writing to learn gives you an edge.

First question

What is an equivalent representation in base 7?

$$(1304)_{10} = (\qquad)_7$$

Before reading further, pause. Think through what you might already know or be able to realize about this statement about the integer 1304.

Take time to find a piece of paper and a pencil or pen. Internalize what is written above and ***without looking into the next pages*** write the same thing on your paper. Then fill in the empty space.

Hints:

1) On the left hand side of the '=' mark is a representation of *positional notation.*

2) You could use an online search engine or Khan Academy to get the sense of a rich and long history regarding positional notation.

3) The base 7 digit set is the set {0,1,2,3,4,5,6}

Let us ***not*** imagine there is nothing here to learn.

Circle of Mentors

It is up to you to communicate to mentors, through your handwriting, into the second school. Qualified individuals will answer your questions and make comments about your work.

Second school communication involves digital message exchanges, scanned to pdf files.

We could put a question mark inside the round parentheses.

$$(1304)_{10} = (\ ?\ )_7$$

Ok, so we are introducing new notation. This notation is necessary if we are to write and communicate using handwritten notes.

Mentors know how to advise you but without knowing you as an individual. Mentors work with profiles and algorithms.

The richness of handwriting is part of how mentors evaluate messages, and how with the help of algorithms and second school training, a mentor will communicate with learners.

Assignments

Most often communication will be in the form of an assignment.

Completing an assignment will require your producing a new handwritten message and forwarding that message into the social learning media.

Note that most of the work will be made through your effort. This is a new model for mentoring and teaching. You learn how to take personal responsibility rather than looking for a way to find help in avoiding this responsibility.

Nature of Perception

It is helpful if you understand what deep learning methods are and how they arise from computer science and behavioral neurology.

A model about human perceptual cycles is used.[2] This cycle involves internalizations of topics found within various curriculums. For example, high school or college preparatory mathematics, or freshman mathematics at a college or university.

Deep learning methods develop your capacity to communicate clearly through handwritten descriptions about topics. These descriptions are then scanned and sent into a social learning media platform, where second school certified mentors read them and, based on our understanding of the material and on our catalog of learning types, a recommendation is sent back to you, via email.

The use of learning media is not something that one does and then stops doing when the next test is passed. Deep learning opens new doors to lifelong learning habits in all areas. Its adoption is therefore correlated with a social movement.

---

[2] More on this is discussed in technical papers available at educationworlds.com.

Who can enroll?

Learning media is designed to change how we see ourselves. Being human, we use iteration to shape our memory systems.

The key tool is a list, one that you make up and change. Repeated internalization/communication cycles create conditions that change how we see our learning experience.

Enrollment depends on understanding rules, context and policies that support this change.

Being a part of the second school means that you will dedicate time to achieve your goals. Iteration develops as you separate your time on task into short study periods. Study habits change.

Our starting assumption is that you are:

A) Preparing to take a college level math course, or
B) In high school and wish to open the door to higher mathematics, for you, or
C) Are currently enrolled in a non-STEM college-level math class.

If this is true, then learning media is right for you.

We cannot hope to change the world,

except in how we change ourselves.

Principle regarding questions

Education is often thought of as:

Teachers posing questions and students knowing or not knowing the answer.

When this is our educational philosophy, biological harm is created in several ways. My statement is conjectural, of course. But bear with me regarding why understanding this harm is part of a proper understanding about why our education system closes the door to higher mathematics for most American children.

Our common cultural disdain towards higher mathematics normalizes a wide spectrum of harms. High stress testing increases this harm.

How is harm persisted in us? There are answers to this question, but these answers are not as yet part of mainstream education literature.

Some elements of a catalog of learning types are associated with common barriers to learning caused by past experiences in math class. The harm can be reduced by using deep learning methods.

It is this simple.

Reflect

Not everyone is affected by the same circumstances in the same way.

One person might not be affected by bad habits expressed consistently by a schoolteacher or college professor. For example, someone might have his or her life negatively shaped by math class. Someone else might not feel this harm.

Realize that each principle we expose about deep learning methods is interpreted in context, created as part of being you. We learn by listening and reading. We learn by writing.

An automated profile development process will discover, with your help, which methods work for you. You must start the process. A handwritten note is how you start. Creating these notes changes everything.

Most incoming freshman students have lost the ability to look at simple math concepts and express questions.

All too often, past experiences in math class inform us that "the system" is only interested in correct answers. How the answers are obtained is not as important as getting credit for them. These past experiences continue reinforcement and conditioning that leads to negative results.

How to ask yourself questions

Often mathematics works with abstractions. The best instructors express abstractions to freshmen students by writing on a white board or chalkboard. Here is one such abstraction:

Let $b = 7$

then $(415)_7 = 4(7)^2 + 1(7)^1 + 5(7)^0$

If b is not specified then

$(415)_b = 4(b)^2 + 1(b)^1 + 5(b)^0$

as long as the digits 4, 1, and 5 are in base b digit set.

$$D_b = \{0, 1, 2, \ldots, b-1\}$$

Examples

$$D_{10} = \{0, 1, 2, 3, \ldots, 9\}$$

$$D_7 = \{0, 1, 2, 3, 4, 5, 6\}$$

"..." means "continue until" so

$$D_{10} = \{0, 1, 2, 3, 4, 5, 6, 7, 8, 9\}.$$

The abstraction is in how *b* is treated. A general statement is made regardless of whether $b = 5$ or $b = 7$.

Within scope

Any question should be "in scope."

Sense of scope and *context* are similar, but we see scope as something that we may easily change. Deep learning methods provide means through which how we form scope changes.

Scope here could mean the meaning of b in the set

$$\{2,3,4,5,6,7,8,9,10\}.$$

If we select b to be 4, then the scope changes to this context.

In which case, for each b we have a different digit set, the D_b.

$$D_5 = \{ 0, 1, 2, 3, 4 \}$$

$$D_{10} = \{ 0, 1, 2, 3, 4, 5, 6, 7, 8, 9 \}$$

$$D_7 = \{ 0, 1, 2, 3, 4, 5, 6 \}$$

Deep learning occurs best when you are making up your own questions.

Finding deep

We are not "studying for the next test." That type of behavior is likely to lead to *shallow learning*, something that we will give a more neurological meaning to later on.

Before we go on, you need to make up two examples:

Examples:

(1) $(461)_8 \rightarrow (\ ?\ )_{10}$

(2) $(3421)_{10} \rightarrow (\quad\quad)_5$

Wait, these are not examples you've created!

It may be easy for you to create original examples like the two above. How to create your own is what we are learning.

Deep learning methods apply to all of life, and how we situate ourselves with what we have.

We are at a critical point. Complicated factors play as you internalize what might seem simple; e.g., make up your own example similar to the examples in the two handwritten images above.

Your unique Self

Each person is different. However, because of the way math class is experienced; many individuals are puzzled as to how to respond. The math class has become well-defined.

How each entering freshman regards this experience is described in terms of a language about various forms of experience.

The degree to which the puzzle stops you is something that tells you about what methods will be most useful to you.

When you communicate to the mentor circle using handwritten narratives, we are able to give you recommendations and explain your learning type.

The second school encodes data in the form of individualizing profiles. And deep learning methods come with a certain type of logic.

You are asked to relax. Find yourself asking questions. The first goal is to understand how to communicate and interact through social learning media.

Use a pencil or pen to write out your own examples. Then find a way to scan this paper into a pdf file and send to the Circle of Mentors.

Strengthening fractional arithmetic

Our purpose is to develop and communicate about arithmetic in a setting that is strange and unfamiliar.

The process of becoming familiar with something is the essence of a deep lesson about arithmetic.

Pausing and reflecting is part of this process. *We are not after a skill, but rather an understanding.* The skill we develop is arithmetic skill.

Arithmetic skill is useful in life, as well as when one is taking a freshman math class. Members of the second school use topic trajectories to communicate with a mentor circle.

Data analytics helps you represent for yourself location within a topic representation of an enumerated curriculum, for example: polynomial functions, their derivatives and anti-derivatives.

You will develop habits such as writing down a new exercise that you yourself will have created.[3] You will blog about how you feel in general, about math class, about deep learning methods.

---

[3] This is "separating" your study time into short intervals.

Being Out of Scope

You are *out of scope* (OoS) when the topics in your study are not what you are expressing. Be careful, this is a crucial point. Some of your expressions in the second school will be responded to by the simple message:

"OoS. Refocus your scope."

How? Return to the basics, list word phrases or words that allow you to discuss number bases. Work within the four-page definition of a trajectory starting on page 35.

When you master the ability to refocus scope, you then request help on the topics in the curriculum you are required to learn by the college or university.

For each element in the new list, ask a question and then write a detailed description about the question.

Be in scope.

Data profiles

If social learning media data is accumulating enough past examples of private messages then a recommendation will be given. So what is going on? Are we using artificial intelligence or natural intelligence?

To address these questions, it is of value to understand a little about what a great behavioral scientist, James J. Gibson, called *action-perception cycles.*[4]

Within second school founding theory, we use complex systems theory to talk about how experience is encoded into memory. Anticipation plays a role, as well as a limited perception given to us by our Mind.

Gibson's work in the early 1950s developed into a school of psychological thought.[5] This school of thought provides mentors with language to talk about how we respond to learning experiences.

---

[4] https://en.wikipedia.org/wiki/James_J._Gibson

[5] Centered at University of Connecticut, the school of thought is called "ecological physics;" it is a subset of complex systems theory.

Ecological physics assumes the mind arises from the physical world, and yet the physical world is subject to decision-making. Decisions arise in ways that do not directly map to physical reality, at least not in a simple stimulus-response fashion.

Social Learning Media

Encoded into who we are, memory shapes our perceptions. Social learning media is based on similar principles. Learning media helps you learn about yourself via how you communicate.

While you are enrolled in a specific course of study, a profile management system is made available to you in various ways. One of these is via a decision-making aid used by human mentors in producing recommendations.

Recommendations follow engaged assessment to produce a temporal map of individual learning. This map is represented computationally in such a way that comparisons are made to a set of temporal maps abstracted into categories and de-individualized.

Learning media uses ontological representation in the form of encoded data structures.

Rediscover natural curiosity

As we first work with deep learning methods, we re-discover a natural curiosity about arithmetic. This is the only goal; when achieved, it opens several new doors. Just having arithmetic skills helps to pass standardized tests.

Re-discovery will necessarily evoke good and negative experiences in math class, or experiences somehow related to your perception about math.

We are each different and yet also similar.

The challenge is to develop a type of internal therapy regarding your feeling of disdain, or other feelings that are natural outcomes from how you perceive your experiences.

For example, you might procrastinate preparation for the next test, miss classes, and then worry continuously. If so, you must decide to change.

We will start with un-familiar concepts regarding counting, not in base ten, but in some other base. As we move along, we shift our focus and establish scope around core topics in the freshman mathematics courses. Our work includes topics about functions, domains and ranges; solution sets and replacement sets.

We increase our ability to internalize un-familiar concepts. Express from internalization your messages into a learning media platform.

Profile building

You must develop an interest, and capacity, to look at concepts from outside the context of standard college curriculums. Surface concepts are powerful, but hidden topics often keep us from understanding these surface concepts.

Without knowing that topics are hidden, you will often fail to understand the relevance of lectures.

Your profile might at one point indicate an interest in something that you did not know was there. Then mentors' recommendations will come. Responding to recommendations produces new messages from you to mentors.

Sharing mastery

The social learning media platform will ask you to look at hidden topics. Because of how trajectories are formed, the mentors are aware of your scope, and will use various means to move you along a path.

Questions you create and share, and questions we show to you, will open your mind. When the mind is open, you will direct private inquiry into something new, merely for the sake of strengthening your skill at arithmetic, and your analytic skills.

Questions may become challenging beyond your expectations.

Consider

$$\left(\frac{1}{4}\right)_6 + \left(\frac{1}{5}\right)_6 = (\,?\,)_6$$

Pause. Internalize the best you can this question. Write what you think is written.

Now let go of the question. Just let it go. The question will return to mind.

Separate the times you reflect on your questions. Make notes about how you feel. Create a blog.

Notation

Let us return to arithmetic and to notation needed to talk clearly about positional notation, and transformations between the number bases.

Let *b* be a base integer. Why does *b* need to be positive and greater than 1?

What is the digit set, D_b, for base *b*?

$1^n = 1$ for $n \in \mathbb{Z}^+$ so 1 cannot be a number base.

$\mathbb{Z}^+ \doteq \{1, 2, 3, 4, \ldots\}$

the symbol means more than "=".
"$\doteq$" means "is defined as".

For some reason, learning how to write out something like what is written above is not often emphasized in high school.

New arithmetical systems

Some have been lucky to get a teacher that did focus on writing.

Perhaps you have seen something like the following written on a chalkboard.

Illustration 1 $D_6 = \{0, 1, 2, 3, 4, 5\}$

Illustration 2 $D_2 = \{0, 1\}$

Illustration 3 $D_{10} = \{0, 1, 2, 3, 4, 5, 6, 7, 8, 9\}$

What is the digit set for $b = 12$?

Can you make up symbols so that you have an answer?

We will work only with bases: 2, 3, 4, 5, 6, 7, 8, 9, and 10, so that we do not create new symbols.

Gearing in

The word-phrase *gearing in* comes from philosophical traditions like phenomenology. When we "gear in," we establish context and scope. There is a shift in our sense of coherence; e.g., what makes sense in a specific context.

Perhaps in the past you have started to gear in to a personal study of some part of mathematics. You may have been enrolled in a math class, or perhaps not. It was just that period of time when you were actively interested in something.

Unless you are lucky, gearing in to math class will be unusual. We each encounter at least one bad math teacher. An extended discussion about what "bad" means, and what are the origins of "bad", are beyond the scope of our manual.

Again, we are each unique; and you may be a lucky one. In which case, you likely are already deep into the arithmetic in bases other than base ten; then again, maybe not.

Enrollment into a learning media platform is a process involving one or more polls. These provide us a first approximation to learning styles. Communication becomes individualized.

Failure in math class has less to do with you and more to do with how teachers are trained. A longer discussion about teacher training is needed. This discussion sometimes develops as part of your subscription to learning media. But we often will wish to avoid this discussion as being unproductive.

We worked on a reasonable business model for the second school, and decided that registration should be free, except for requirements that polls be completed. This manual is provided at low cost from Amazon. It is one way to introduce you to learning media.

After a time, mentors expect a first topic trajectory to be placed into your drop box. A few of these handwritten narratives will be responded to without compensation for enrollment into our platform, so that you may get a sense of what to expect.

Then a subscription fee will be established. Each person is evaluated as a person. The fee is arranged to help support the media.

References to a textbook

It used to be that college algebra was the only mathematics course for non-STEM majors. Now we have many variations and extensions.

If you have a textbook, and we have a copy of the same book, we are set to begin developing trajectories as part of your enrollment in social learning media. This is not always necessary, but is the most used option for study with social learning media.

A parallel study is when an individual is enrolled both in a college non-STEM mathematics course and in a learning media environment.

You first learn how to write about topics in your course syllabus. This generally requires a face-to-face workshop lasting three weeks, with about 3 hours a week in a classroom. But there are exceptions. We can do a single workshop in two three-hour sessions on a weekend.

We then learn how to establish scope by selecting part, but not all, of the curriculum defined in the university's standardized syllabus. This task is made interesting because you must make decisions. In making decisions you reveal how you feel.

Return to arithmetic and notation

Scope was identified as arithmetic in a number base not equal to 10.

Review:

$$(314)_{10} = (\ ? \ )_6$$

Two ways to find the answer are:

1) make a list of powers of 6

$6^0 = 1$
$6^1 = 6$
$6^2 = 36$
$6^3 = 216$
$6^4 > 314$

Now divide 216 into 314
1 remainder 98
$99 \div 36 = 2$
remainder = 28
$27 \div 6 = 4$
remainder = 2

So $(314)_{10} = (1242)_6$

2) write $(314)_{10}$ in positional notation

$3(10)^2 + 1(10) + 4$

express all numbers as base 6

$3(14)^2 + 1(14) + 4$

because $(14)_6 = (10)_{10}$

now multiply each number using base 6 multiplication.

14 × 14: 104, 14 → 244

244 × 3 → 1220

and ADD

1220 + 244 + 4 (crossed out) — opps

1220 + 14 + 4 = 1242

So $(314)_{10} = (1242)$

Now try to make up your own exercise. You should notice from above that mistakes are easy, but that if two pathways are used, the answer will have to be the same – no matter what exercise you make up. So we do not need a book.

First assignment

A first assignment is to create a four-page topic trajectory, scan this to a pdf file, and place this file into your second school drop box.

Work on creating simple arithmetic exercises during short periods of study in between your study of your class textbook. You will begin to see what scope is and how to identify various parts of the curriculum. Scope is identified by the actual list of topics. You write this list on the first page of your trajectory. This list establishes context and scope.

You should continue to post trajectories about number bases and arithmetic. At one point, you will quickly create four-page trajectories describing topics from your college curriculum. Knowing how to enumerate all of the topics in your curriculum is part of our objective.

You might choose to spend a day or so, at random times, developing a single exercise. You make up these exercises, or you find them in a textbook.

Check your answer by developing detail in different ways. If quality is high, the post will be "de-individualized" and posted (with your permission and without identifying data).

Glass Bead Game[6]

High quality posts will be viewable by others also enrolled in the second school, but again no identifying data will be seen by anyone.

It is like going to the gym to exercise. You are just exercising that part of your brain that does arithmetic. Simple review of skills you already have increases your capacity to learn.

So set this book down and pick it back up tomorrow or the next day.

Go online and develop an emerging version of the glass bead game. Through your work within the second school you will see simulation worlds-like 3D avatars represent your trajectory location and status.

---

[6] The reference here is to the historical literature about an ancient game that combined I-Ching and Chess.

Making a list

Can you list a set of topic names within a specific scope?

What about a recent assignment?

Having experienced working for short periods at random times for a day or two, you may have developed an internal feeling about what is going on.

This is what you want to communicate into social learning media.

Make a list and then create a trajectory. Send this into the social learning media platform. Did you see the need for a multiplication table and addition table in the base you are working with?

Consider
$$\begin{array}{r} 327 \\ * \ 45 \\ \hline \end{array} \text{ in base } 8$$

$$\begin{array}{r} 451 \\ + \ 52 \\ \hline \end{array} \text{ in base } 6$$

$$\begin{array}{r} 637 \\ 45 \\ + \ 24 \\ \hline \end{array} \text{ base } 9$$

Did your mind open up into these types of questions? If so, communicate this. If not, communicate this.

See if you follow the next steps:

Notice that $(327)_8 = (3(8^2)+2(8)+7)_{10}$
$= (3(64)+16+7)_{10}$
$= (192+16+7)_{10} = (215)_{10}$

and $(45)_8 = 4(8)+5 = (37)_{10}$

so $\begin{array}{r} 215 \\ *\ 37 \\ \hline \end{array}$ in base 10 will be a base 10 number that when converted to base 8 is the same as $(327)_8 * (45)_8$.

The notation is vital to understanding.

If you are unsure, you may develop a trajectory about what you see and what you do not see. Send it in.

The following page is a sample trajectory about number base arithmetic.

Trajectory example

Even if you are not enrolled in social learning media, see what you can do to find natural things that can be done here.

Example trajectory:

topics
positional notation
digit set
number base
multiplication
addition

comfortable
digit set
number base

uncomfortable
multiplication in base 6
addition in base 6
positional notation.

detailed description of
digit set + number base

*	0	1	2	3	4	5
0						
1						
2						
3						
4						
5						

+	0	1	2
0			
1			
2			

blog entry.

In some cases, this is easy. But you might struggle a bit here, and this is really expected. First the scope of the trajectory has to have been internalized.

Rely more on yourself

Did you spend a day or so making up exercises at random times? If you did not, then put this book down and look to yourself for next steps.

Location in Representation Space.

Let me continue into the Foundations topics.

Consider the addition + multiplication tables for base = 4

+	0	1	2	3
0	0	1	2	3
1	1	2	3	10
2	2	3	10	11
3	3	10	11	12

*	0	1	2	3
0	0	0	0	0
1	0	1	2	3
2	0	2	10	12
3	0	3	12	21

Now what is $(98)_{10} \rightarrow (\ \ )_4$?

What is 32 * 13 in base 4.

$$(32)_4 * (13)_4 = (1202)_4$$

$$21(22) + 20$$

$$9(10) + 8$$

$$3(4) + 2 \qquad 4 + 3$$

$$14 * 7 = 98$$

How you manage work effort is indicative of how you feel about math, and about yourself.

Be aware that change is occurring and that learning is being demonstrated.

***As your profile changes,
your mentor avatar also changes.***

Your mentor avatar is your reflective self.

Comparison to traditional teaching

Traditional teaching tells students what to learn, when to learn, and how to learn.

This is a mistake.

Biological systems learn all the time and in unique ways. When some authority tells us what, when and how, then there are natural pushbacks. This is where the harm comes from.

But we are here to undo damage, if any. Yes?

Deep learning methods open the door to natural habits. Then the mind is open to learning.

Let us review where we have been, already.

Example: You choose a base from the set

{ 2, 3 ,4 ,5 ,6 ,7 ,8 , 9, 10 }

and construct the * and + tables.

Hint: Move things around, not just so that you are following someone else's rules.

Consider the exercise:

$$(354)_7 \longrightarrow (\ ?\ )_5$$

What do you need?

Assignment example

Are there at least two distinct ways to find the answer? Of course, but why find two ways to develop the answer when only the answer is needed.

What an odd question.

After reviewing any example, set aside this book.

After a while, find yourself, with 15 minutes of time, some unlined copy paper and a pencil or pen.

Construct a new four-page trajectory. The closer you follow formatting rules, the easier it is for a certified mentor to respond with a recommendation.

AI plus human-in-the-loop

We live in a remarkable time. Having a set of enumerated topics helps our assessment-recommender response process. This capacity changes everything.

Engage in study and do more. With social learning media you co-develop a unique profile. Your profile creates an avatar and this avatar interacts with other avatars in a glass bead game.

Algorithms make part of the response to your trajectory. You should have information about how this works. Obviously this information is complicated.

You send in a four-page trajectory. Second school software separates the four pages into parts. Algorithms create a referential map to a topic map.

Topic maps are formal structures that represent the curriculum you have selected within scope. The topic list tells us where you are located, in one sense. But we human beings are not that simple.

Your personal blog

Your blog is read in two ways. First an approximate profile was made based on your enrollment polls. This profile is changed as new information is processed.

Character recognition converts some of your handwriting to ASCII and this data is stored in a file associated with your account.

Lastly the profile is given to a certified mentor. He or she will not know your name. The Circle of Mentors will have AI-aided analytics to guide his or her recommendation to you

Your personal blog is under your control. You may share with friends in a way similar to how Facebook® *share* works.

Your blog will not be shared at all with anyone not listed in your share file.

Second Directive: You may make public your blog only if shared information is completely de-individualized.

Prime Directive: No criticism of the first school is allowed. This would be OoS.

Challenge topic:

Intersection between a line and a parabola

Have you ever wondered about this topic? Perhaps you already know how to do this using a graphing calculator? You may know how to graph parabolas and lines without a digital aid. If you do, you likely had no problem passing your last math class.

Let us address a difficulty found by so many. The scope of what is supposed to be tested on is not clear. Finding location within a part of the standard curriculum is the way to reduce that difficulty.

It is entirely likely that you have been in a math class where lines and parabolas were in scope.

When you were attending class, which might not have been 100% of the time, core concepts were discussed. These included the two most essential concepts of a

replacement set and solution set.

The way we discuss replacement and solution sets requires an extension of notation.

Writing the notation

Many notational elements used by professors in non-STEM math courses are not in books. Some college courses are taught now with PowerPoint presentations. This is a huge mistake.

This is an old problem with a specific technical barrier. The keyboard has no way to place proper symbols into a text file.

Members of the second school use handwritten message exchanges to communicate about even the most basic concepts in high school algebra. Once you start to do this, learning is accelerated.

You are able to ask a second school mentor avatar questions that have been hidden from your perception.

Class board notation is often not used except by mathematicians. So of course some college classes will use this notation as if everyone understood and had seen these symbols.

Independent use of social learning media allows the type of free association necessary to build a sense of location within the topic field defined by an enumeration of topics from standardized curriculums.

Why use novelty

We use novelty in social learning media when mentors feel that a transition is possible. We imagine that individual behavior may be changed in slight ways, and in some cases a small change makes a big difference.

We are looking for shifts in your view about the learnability of curriculum you are studying.

Aware processing of novelty involves very different pathways in the brain than does our processing awareness with familiarity.

Novelty orients us in a different way, at least for a little while. It is during this time that our mentors expect transitions in our behavioral expression.

The theory under social learning media suggests that for each of us, we have a small set of behavioral generators. These work like the phoneme generators of voice. Perhaps fewer than 50 reflex arcs are used by the neural system to initiate motor programs involved in your speech.

Deep learning occurs when one or more of these programs become fundamentally altered. Novelty opens the door to the production of new memory engrams. These then may eventually develop interactions with your expressive behavior.

Discussion about intersections between solution sets

We will take a break here and look at an illustration of what a trajectory looks like.

We will then return to our work on arithmetic in bases other than base ten.

List of Topics:

- 13.1: Antiderivatives & Indefinite Integrals
- 13.2: Integration by Substitution
- 13.3: Differential Equations: Growth & Decay
- 13.4: The Definite Integral
- 13.5: The Fundamental Theorem of Calculus
- 14.1: Area between Curves
- 15.1: Functions of Several Variables
- 15.2: Partial Derivatives

A first listing of topics

The student is in the last part of business and economics calculus. The list is made without classifying topics as *comfortable* or *un-comfortable.*

Separating a list into two lists

Understood:	Misunderstood:
· 13.1: Antiderivatives & Indefinite Integrals	>13.3: Differential Equations: Growth & Decay
13.2: Integration by Substitution	> 14.1: Area between Curves
13.4: The Definite Integral	>15.1: Functions of Several Variables
·13.5: The Fundamental Theorem of Calculus	>15.2: Partial Derivatives

The previous list is then used to separate topics into those comfortable with and those not.

We see, very quickly, that a recommendation might be to establish scope around any one of the four in the second list. 15.1 and 15.2 (referring to a student's text book sections) are very different from the first two, and each of these is different.

So what will be the recommendation?

Family of solutions and particular solution

Look at detailed descriptions of anti-derivatives. We see difference between the concept of a family of solutions and a particular solution. This difference is communicated in the detailed description shown below.[7]

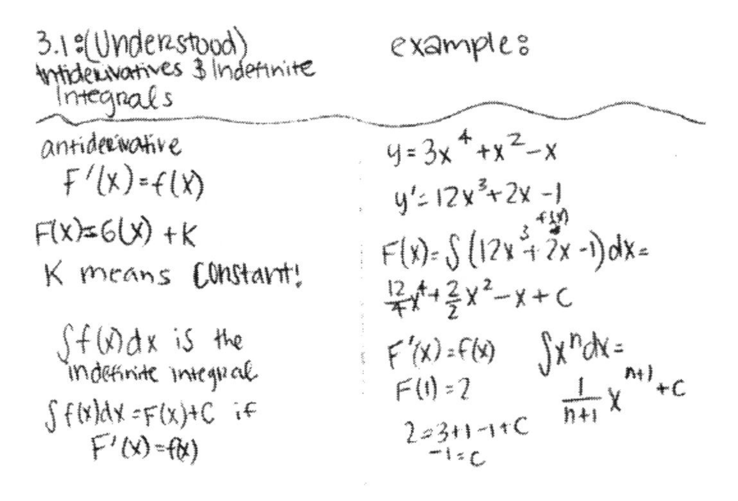

Note that the description is written well, "equal" symbols are properly used, and there is clarity in this expression. It is a bit cramped towards the end of the description.

---

[7] The family has a parameter *C*. A particular solution requires the specification of one point in the plane. This is found to be -1.

Continued illustration

As trajectories are reviewed, new profile elements are inferred and these modify your avatar.

The following illustration involves anti-derivatives of polynomials.

EXPLANATION:

The properties and formula rules were easier to follow because they were explained clearly.

We recommend that more space be used towards the end of the exercise so that the several concepts being referred to are better separated.

If one now studies 13.3, differential equations, growth and decay, one finds the *Exponential Growth Law*. This law requires two points to be known rather than just one, and therefore is different from what the student has detailed. The recommendation is to study this difference.

Continued illustration

By looking at the image below, we see that writing is being cramped. There are stylistic concerns that limit what may be communicated.

3.2: (Understood) Integration by Substitution

Reversing the Chain Rule

$\frac{d}{dx} f[g(x)] = f'[g(x)]g'(x)$

differential dx is for x

differential dy is for y

So, $dy = f'(x)dx$

If $f(x) = x^3$, then

$dy = f'(x)dx$

$= 3xdx$

example:

$P'(x) = .015e^{-.01x}$

$P(x) = x^2 + 1$

$\int P'(x) = \frac{1}{-.01} \int -.015e^{\frac{.01x}{-.01}} dx$

$P(x) = \frac{-.015}{-.01} \int e^{t} du$

$u = .01$

$du = .01dx$

$= \frac{.015}{.01} e^{-.01x} + C$

Several concepts get a little mixed up

Continued illustration

EXPLANATION:

As long as you follow the book this lesson was simple.

Here we see overconfidence, but perhaps not something to make a big deal with. Clearly there is a sense of accomplishment.

Continued illustration

Where I am in this class:

I think certain topics are harder to understand and get correct compared to others.

Blog entry reveals a lot about this student's sense of location

What I understand during class:

I feel certain topics are too hard to understand at once. So, going over them one section at a time is helpful.

Interesting insights

Continued illustration

My feelings:

I get confused at times. The book helps to understand the material better and the examples given in class for 13.1 and 13.2 were good to me. I need to Review a lot more.

Submitting your work

You do not always need to submit trajectories. You may just develop your work and keep most of it. We ask that when storing your trajectories you fold them lengthwise. Folding this way helps when you scan to pdf later on. Lined paper is like window blinds, they obscure the view.

Or you may just use a three-ring notebook, and get one of those hole-punching tools. A lot of this depends on what you feel comfortable doing.

Folding separates an inside from an outside. Your name should only appear on the outside. As we process an incoming trajectory we want to use algorithms without your name being on the paper.

To be a member of the second school, you must agree to our privacy statement. Images with names will not be input into the algorithms.

Social learning media will have virtual portals in coffee shops and teahouses. You may find a place where scanning to pdf and email is supported. You may also find self-organized groups meeting at various places on and around campus.

Social learning media is completely separate from university, college or school records. We do not discuss habits or behaviors of individual faculty members, as this is OoS (Out of Scope).

Creativity renewed

A core purpose here is to open up creativity. Do you see mathematics as a creative process you participate in? Openness means open to seeing what is simple in this important cultural legacy.

Decide how you want to produce and maintain learning trajectories. Change the rules so that they fit you. Communicate your location.

In the previous illustrations we saw one person using one trajectory to find location. Another person might struggle with some of the same things. Recommendations could be about narrowing the scope. Yet narrowing scope is sometimes meet with frustration.

There is often not enough time left to save the semester. We must have time to rescue an individual who is in danger of failing a class or not passing entrance exams.

Experience tells us that two weeks is just long enough to engage fully in the methods. This may be done without meeting face-to-face. Small group meetings are offered.

Small group meetings are established anywhere where several individuals are brought together. Registration is the first step towards creating a small support group.

Return to positional notation

Let us return to arithmetic and notation. The previous images of handwritten work were within a specific scope found during the last part of a two semester freshman requirement for economic or business majors.

Let *b* be a base integer from the set:

$$\{ 2, 3, 4, 5, 6, 7, 8, 9, 10 \}.$$

Why is *b* positive and greater than 1?

Pause.

Go do something else, like wash clothes or something. Allow your full mind to work as you engage in anything other than this question.

Does the question come back to you? Do you see shadows of change in how the question feels?

What is the digit set, D_b, for base *b*?

V

$1^n = 1$ for $n \in Z^+$ so 1 cannot be a number base.

$$Z^+ \doteq \{1, 2, 3, 4, \ldots\}$$

the symbol means more than "=".
"$\doteq$" means "is defined as".

What is the digit set of $b = 12$?

We will work only with bases 2 through 10 so that we do not create new symbols.

Illustration 1 $D_6 = \{0, 1, 2, 3, 4, 5\}$

Illustration 2 $D_2 = \{0, 1\}$

Illustration 3 $D_{10} = \{0, 1, 2, 3, 4, 5, 6, 7, 8, 9\}$

As you see, handwriting is not always as easy to read as typed text. But then again, the ability to write outside of the lines is important. We also see various psychological and learning type variations by simply noticing how the strokes form letters and other symbols.

It is the use of these other symbols that trajectories require.

Insightful mentorship requires communication. A vetting process model is used. Your communication using trajectories is essential as input to that vetting model.

Several questions have been asked, and you are to consider developing properly formatted topic trajectories having scope defined by those questions.

Are you ready?

Part Two: The Method

Who this book is for?

As you work with deep learning methods your perspectives will change. In particular, your perspective towards the study of college level mathematics will deepen. Where there are now feelings of regret, or even disdain, your perceptions change.

Your mind opens up not because of a teacher or a tutor, but because you have taken charge. Finding location is essential, and a vetting process within the social learning media aids:

"finding self as located."

A grass roots movement is changing how mathematics is taught and learned. Support is available online via our web sites. From there you may join a social media and/or find links into other parts of the second school social movement. You may even start a local branch of the second school network.

Much of the value of deep learning is communicated within social media. As such we do not have physical boundaries. The social movement will grow.

It is good to have a portal into social learning media from a coffee shop or other gathering place. This is something you are encouraged to establish after your first workshop.

The traditional classroom and Second School

In what follows, we think critically yet constructively about how math class is conceived and administered. This is Out of Scope (OoS) but necessary to overcome high levels of pretense from students, teachers and administrators alike.

Instructors in math class often earn professional success through traditional practice. But sometimes the relationships between instructors and students are disconnected. Second school practice addresses at least some of the causes of these disconnects.

With second school methodology, you are able to focus on what your barriers are. Specifically, your perception about you and math class. A profile is built with humans and algorithms in an *action-perception cycle.*

We will discuss this more fully later and in face-to-face workshops, particularly during introductory sessions.

All college curriculums are well-specified. Each college has a specific syllabus for non-STEM major course requirements. We simply represent each specific syllabus as a topic map.[8] With topic maps, your cognitive location is described as a location within a space.

---

[8] A topic map is a computable representation of things and relationships between things.

What problems do the methods address?

Second school processes and techniques create a representation of where you are "located" in your traversal of what is represented to you as a "forest" of topics.

This location is not where it could, or should, be. Our philosophy is that experiences in your past are limiting what you can do in the present.

Learning media is designed to overcome this type of limitation. Learn to move in and around the topic representation spaces. Excel in math class.

Social systems are embedded in history and in interactions and constraints of the moment. In some cases, parts of a system evolve and get stuck in what a mathematician would call a "basin of attraction." We also have mindsets that get stuck.

But being stuck with a sense of disdain about math class is not something you have to put up with. We as a society do not have to put up with it either. Find your location within the field of topics. Once found you are able to move about that location.

More on location

When a natural system's behavior is stuck, we might recognize failure but might not be able do anything about the origins of that failure.

Our education system is stuck in a local basin of attraction, and in varying degrees you are also.

The single most important factor in not graduating from high school is high school algebra. The single most important factor in not graduating with a two-year degree is the college algebra requirement. So this is a big deal.

We believe that math education in school might be fixed, but our educational systems have too many structural challenges. This will change as social media evolves into platforms dedicated to individualized education.

So what will this be like, this world of the future?

Unintended consequences

We observe progress towards a liberally educated America. But this progress is inhibited by hidden causes and complex social histories.

Deep learning methods work to recognize and overcome obstacles created by a less than perfect system.

Deep Learning Manuals, the series, might be a strong sign of what is to come. A transformation of our society is to arise from a resolution of the crisis in American education practices.

Advanced algorithms embedded in social learning media might guide this transformation.

We will see.

Independence

We predict that students will use social media to learn far more than what is now expected. For this to happen, social media has to be separated from the control of the current system.

This is what learning media does.

Deep learning methods change behavior and viewpoints, including behavior and viewpoints expressed through instructors. The task is very difficult, in part because the individual must recognize how to be different when the system is not.

We cannot easily change when we are a product of a system as powerful as our educational system. But suppose that there is a new system, one that does not challenge the old one, and yet creates new opportunity?

Motivation

You'll probably agree that most persons taking non-STEM mathematics courses are not exactly enjoying the experience. One young lady in Austin told the story of being in a calculus class of 220 students. The professor was not speaking clear English, and all of the student teaching assistants had the same problem.

Because she failed, the young lady changed her major from biology to finance. This type of story is told over and over again. This type of experience is common.

There are important exceptions, of course. Each individual is asked to see his or her experiences in context. Most will feel unable to re-contextualize the study of basic concepts, in part because the system's expectations are as they are.

Can you lead the way, and show others? Can you make your college math class an enjoyable and rewarding experience?

Deep learning methods are designed to alter the underlying belief structure that is part of who you perceive yourself to be. In this sense, the methods help you identify your individuality as no longer blocked by hidden expectation.

Freedom!

Common experience

Our experiences in K-12 math class shape how we view learning. This is true, in particular, when learning college-level mathematics, whether in a classroom or online.

We do not start a math class as a blank slate. Nor are our past experiences always purely positive. Through a vetting model, learning media produces solid evidence concerning your location in representation spaces.

As you develop an ability to write about basic topics in your math class, you come face-to-face with your past experiences. Send in your trajectories. Mentors and algorithms will recognize specific characteristics.

This is all kept confidential. Various algorithmic methods are used to develop a deep structure-based model of your current perceptions.

Using a universal model of curriculum and learning types, learning media represents changes to your perception over time.

A deep structure model is derived from your trajectories.[9] Then your private data is set aside, to be reviewed only by a certified mentor or computer program.

---

[9] Theory of deep structure, as used in learning media is developed in the publications of Paul Stephen Prueitt.

How the methods work

Second school mentors recommend various tasks, such as working out four or five exercises and then writing out a detailed description of the topics involved.

You become more in control of the present experiences with the math class.

We recognize that most non-STEM majors do not enjoy the required one or two college courses in mathematics. So motivation is a concern.

Motivation comes from finding a comfortable place to be.

Mentors are trained to assist you in learning to write mathematics. This is a simple idea, but sometimes difficult to achieve. Then we move into curriculum spaces that you define for us.

We are assisted by a deep knowledge of the curriculum, and how topics are related to each other. We also see hidden topics that should be in the curriculum but often are not.

For some, the idea of taking class notes and working out the detail of exercises is natural and reasonable. It is agreed that some type of measurement of your viewpoints and perceptions is necessary.

User-defined scope

There are common pathways through representation spaces. When algorithms first process your trajectories, a categorization is made. This does not occur only at one "level of organization." Commonalities noticed across many instances are encoded as a small set of recognition primitives.

Learning methods work by assisting the individual in finding location. After learning how to communicate, we develop an enumeration practice. Your enumerations automatically create parallel representations in a topic space.

From formal means, defined on this space, algorithms create your profile. Formally this abstraction is a computable representation.

You make a list of topics that are in the curriculum. This is a first step in communicating with your teacher about the subject matter he or she is teaching.

Students "find scope" by listing names of topics found in the textbook. Topic maps encode all topics that are covered in the textbook. Mentors use this encoded data to quickly provide a recommendation.

Interaction with mentors produces a *learning trajectory* represented by a simple graph, with nodes being the enumerated topics of that curriculum.

Four Methods

Enumerating and Separating: Deep learning methods ask you to give your own names to a set of topics within "scope." You then separate that list into two: topics with which you are comfortable and topics with which you are not.

Scope: In the establishment of a "universe of discourse" in the context of one's location within a curriculum, scope may be set by a test, the final exam, a homework assignment, or a practice test.

Handwriting about Topics: One cannot type reasonable math class notes. One needs the ability to handwrite clearly. Notational devices we teach are useful. These devices are related to logical expressions, sets, domains, ranges, replacement sets and solution sets.

Blocking and Separating: Like speech acts, mental occurrences involve two separated time scales: one fast and one slow. **Blocking** is like cramming for a test. Fast thinking is enhanced and immediate performance on a standardized test increases. But fast thinking separates surface and deep thinking. **Separating** one's study times into many short duration periods associates fast thinking with slow thinking. Your mind finds a synthesis between fast and slow. This synthesis may now be integrated into your sense of self.

Grounding Deep Learning

Deep learning methods have philosophical and formal grounding in natural science. For example, various research literatures talk about fast thought and slow thought.

Fast thought is generally little more than a guess. With fast learning we tap into shallow social thought. “Why learn this when no one else has learned it?”

Slow thought occurs when the individual’s mind is restful and contemplative.

Why learn math?

It is not all about the system. There are various private reasons for one's inability to learn math. These reasons are explained by behavioral neuroscience. For example, variability of stimulus creates a potential to initiate self-directed processes, when an individual feels "mentally blocked" by a math question.

Deep learning means that the pace with which we move must slow down until a sense of location is found. Deep learning means a lot more, but this sense of location is the single most important key.

In the first part of this manual, foundational topics are presented while looking at topics where confusion and dis-comfort are common. The presentations are two ways, from you to the second school and from the second school to you.

Invention of the Topic Trajectory

The topic trajectory was invented in February 2016. I was addressing criticism from tenured faculty in the Department of Mathematics at Texas State University. Several sleepless nights were spent focusing on how to improve deep learning methods, as they were then defined.[10] [11] [12] [13]

I saw suddenly one night that trajectories are a means through which an individual student clearly tells the instructor his or her location within a curriculum. This "message" might also be automatically encoded in assessment-recommender software.

---

[10] Paul Stephen Prueitt (April 17th 2015) A New Data Source for Design-Based Research on Deep Learning Strategies, presentation to the Mathematics Education Seminar at Texas State University

[11] Paul Stephen Prueitt (Feb. 6th 2015) Deep Structure Architecture for Machines and Humans, presentation to the Mathematics Education Seminar at Texas State University

[12] Paul Stephen Prueitt (February 2015) Deep Architecture for Human and machine Learning, seminar presented at Computer Science Seminar at Texas State University

[13] Prueitt. Paul Stephen (June 23, 2014) Individually Directed Inquiry. R L Moore Legacy Conference 2014, Denver CO.

Finding location

Of course finding location is not always easy. Once this sense of location is found, deep learning may begin. But even then we must move slowly.

Moving slowly is important due to predictable consequences from past learning experiences. These experiences are still part of who we see ourselves as.

For example, experiences learning math are often incomplete. For whatever reasons, a few essential topics are hidden from our view.

Change the game

Our capacity to direct our own inquiry is often restricted by consequences from past experience. For example, unsuccessfully confronted topics create cognitive barriers. Class experiences do not always help. Missing class is also common.

Confusions are accepted as normal. Very little signal is given that you can in fact know mathematical concepts. We just struggle on, not knowing if the course will be passed this time or not. Truth is to be found nowhere, it seems.

Social learning media gives insight about how to re-establish a sense of location within a well-motivated and self-directed inquiry. You communicate what you are comfortable with. Social learning media then reinforces positive learning behaviors.

You learn how to discuss topics you know about but are not comfortable with. The second school is set up to support this type of communication outside of class. We help you more fully participate in class.

Internalize your understanding. Demonstrate skill and create new ability to synthesize. Within a dedicated social media, you communicate with avatars. Learning is separated from your classroom experience and is de-individualized.

Math class

You know what math class is like.

In most cases, college math class involves lectures where a professor writes on a chalk board or a white board. Often what the professor writes is not re-written by students.

In most cases, you observe passively. If attendance is not required, attendance drops to around 60%, unless there is a test.

In our program, notes about topics are written both inside and outside of class. These notes become trajectories that you communicate within the social learning media.

Trajectories are like tweets.

Taking good class notes creates a resource. These resources are sharable between you and your colleagues. The primary function of your trajectory is to communicate where you are to you.

Once a profile of your location is available, recommendations about what to study next are given.

Change what happens in math class.

Social Learning Media

Abstractions about avatar profiles are produced from many trajectories. These are shared among students and mentors. Communication helps build a sense of location.[14]

If asked by an instructor in class, "What are we studying now?" the response is usually silence. Repeated enumeration of topics changes this so that the names of topics are available for discussion.

You will surprise your instructor.

Communicating at this depth requires confidentiality. Anonymity is a key element in how social learning media works.

Unlike the first school, we keep no personal data about you unless you are currently enrolled in a workshop or mini-course. Even then, limited access is provided and only for the purpose of improving your learning process.

Trajectories are communicated to establish a capacity to connect within a social media dedicated only to learning academic subjects.

---

[14] We hope to have published data available in the year 2017.

Second School network technology

We know that certain behavioral types exist, and that a categorization of such types is possible.

From these categorizations we understand better the specific learning process in which you are engaged. We also develop a precise location for you within a topic representation space.

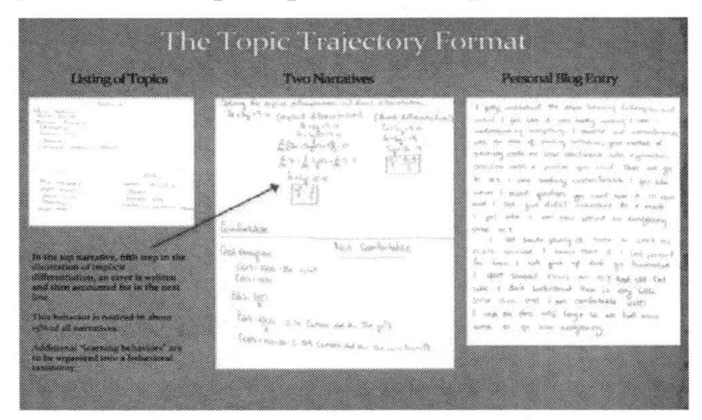

From March 2016 presentation to the Data Science Club at Texas State University

Role of Artificial Intelligence: Algorithms and rubrics code handwritten topic narratives and create learning behavior profiles. These profiles then provide guidance to a recommender system, similar to what is used in Linked-In®, but having specific innovations not yet seen in mainstream social media.

Enumeration of topics

When used successfully, finding a sense of location is the most obvious result of deep learning methods.

In the figure below one might imagine each line to consist of handwritten names of topics as perceived by the individual. An abstract model is presented below.

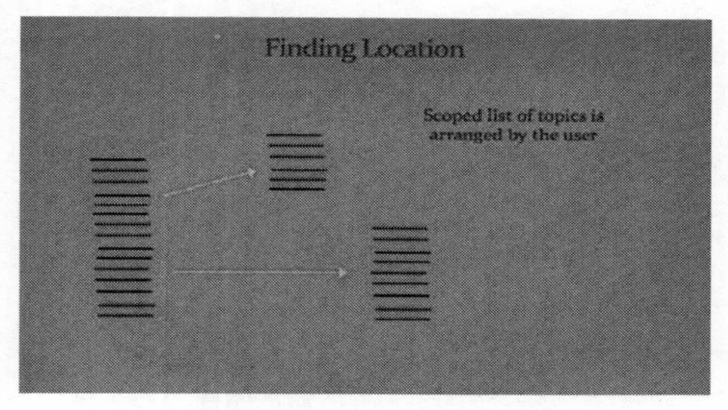

First Page of a Student Trajectory

Enumerating topics is not something that a student in a freshman non-STEM math class expects. There is resistance to doing this. This resistance may be overcome.

Sometimes one must see a learning trajectory as something external to previous experience. In this way we allow ourselves to change our perception about math class.

How difficult is enumeration?

If you have never really felt comfortable with math class, you will likely feel discomfort when attempting to list topics. This discomfort is a significant part of the sense of self for many.

Learning how to write mathematics and how to enumerate topics is a new skill, and not one that has been used very often. Those that are good in math do this, or something similar. But you are different. Is this difference imposed? Might it be removed?

Deep learning methods require that a student enumerate a list of topics and then separate that list into two: comfortable topics, and uncomfortable topics. But enumerating requires an acknowledgement of common private consequences from poor school experiences in math class.

We take a long look at who we see ourselves to be.

Colleges and universities have the same response. It is difficult for the system to actually recognize the decreasing capacity of incoming freshmen with respect to college level study.

Enumeration is the first step

Repeated, daily enumeration releases you from what is essentially a high drama private narrative. You have been involved in drama all of your life. The system is also in a narrative.

New habits are formed. Once you see a topic as an element in a "space," you can search sites such as Khan Academy and the like to get short clear videos about that topic. But there is more value than just using names as search strings.

Social learning media and face-to-face workshops are designed to use enumerations as a means to communicate to you as an individual. This means that you internalize topics and create an ability to work test questions more successfully.

But it also means that you get quick advice on how to study and what to study. You develop specific questions and communicate these questions clearly. Recommendations guide you forward.

A remarkably simple success formula

The key to success for adaptive assessment engines is a representation of location for the individual, and a repertoire of go-to instructional plans to be recommended for that individual.

Plans are created from learning trajectories by mentors. A next assignment is sent to you via email.

It really is that simple. We just have to do it.

Learning environments

Rewards from enumerations come because enumerating into categories defines a formal location and sets parameters used by algorithms and human mentors in the system.

A computable data structure encodes your learning trajectories. You help identify types of learning behavior through your blog entries. Certified mentors read these blog entries.[15]

The three primary learning environments:

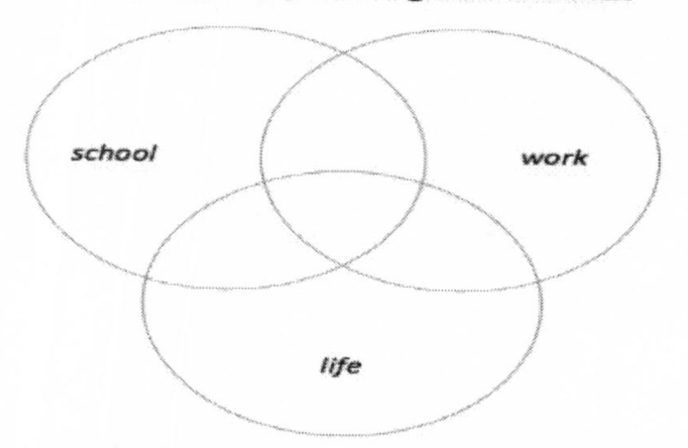

In life, writing involves keeping a journal. All the great thinkers, scientists and inventors kept journals. Einstein said his best teacher was a yellow pad and pencil...

[15] Each is contractually bound to protect your privacy.

Ultimately, we expect to see topic map representations of school, work and life environments.[16] For this to work, we need only your human input and some algorithms. No one else should be involved.

---

[16] Deep Learning Manual: the knowledge explorer's guide to self-discovery in education, work, and life, by Dr. Arthur J Murray (2016)

Portal into Second School Network

The manuals have a written and unwritten part. The unwritten part includes a digital portal into a social learning media. The manuals are low cost and available as e-books from Amazon.

Our social learning media is equipped with unique learning tools. These include adaptive computer-based assessment, recommender systems and Linked-In® type social media analytics.

A new class of advanced data structures is constructed using both human and artificial intelligence.

Digital profiles group individuals in ways that are relevant to the learning process. Each individual may now be matched to specific mentoring services.

Second School Network mentoring services are both face-to-face and online and are designed to be universal, de-personalized, topic-oriented subscription products following deep learning methods as defined by the second school.

Mentors are certified and payment to mentors must come via the Second School Network enrollment process. Training in deep learning methods provides a common set of interaction policies.

First Principle: Study only for short periods

We often count study time as time spent being frustrated. Rather than "study" for an hour, it is better to study six separate times for ten minutes.

In other words, study often, for short periods.

This is called "separating." Separating naturally engages a slower time scale, along with the engramic substructure of your own personal awareness.

It is also better to know names for topics that you have learned, and to be able to make judgments about to what degree this learning is deep learning. Test results should be no surprise.

You communicate with a social learning media via handwritten messages, photographed with your cell phone, or by using a printer or computer, and uploaded to a dedicated digital server. Once the digital images are uploaded they are viewed by a mentor and comments are provided via email.

You separate study times into short segments. Once some topics have been studied, we recommend that you go do something else. Then return to your studies and produce a trajectory. Send this into the social learning media platform.

Within a short period, your mentor will send you a recommendation as to what to study next.

Prerequisites

There are no prerequisites. We assume that your unique experience in math class includes one or more of the many common experiences found in the K-12 math classroom. These experiences often taint perception about higher mathematics.

For example, we often spend time learning test-taking techniques while at the same time diminishing the importance of actually understanding the topics. Cramming for the test is just one of many bad practices.

Deep learning methods set up communication links between you and an entire social network. Your personal identity is distributed into representational spaces, and lost except for very well-defined relationships with mentors. Algorithms help maintain the link and protect your identity.

AI and Human-In-the-Loop processes make an assessment of your ability to evaluate a topic, capacity to synthesize between topics, and demonstration of skill:

{skill; synthesis; evaluativeness; completeness}.

These will become clear as we develop the methods. For our purposes, skill is as we have come to understand from standardized testing.

Self-directed study

Our model details how classroom experiences are sometimes very good, but often are not meeting your needs. To meet true need, one must recognize the nature of problems that stand in the way.

How can one explore what one cannot communicate about? So we learn to write mathematics. Once we have the ability to communicate, then new experiences in math class enhance self-directed behavior.

The question: "What do I need to do to earn an A?" has become shallow. "What do I minimally need to do to earn an A?" is even shallower.

Synthesis and evaluative dimensions are difficult to measure. Knowing what the questions should be requires a different perspective than what we see.

You learn to make evaluations and to synthesize relationships. Certified mentors will help you.

The practice of producing topic trajectories re-creates a sense of control over how we learn.

Three-dimensional Internet

Eventual embedding into a 3D simulation world is anticipated by technological developments in virtual education environments, many of which are derivatives of 1980s collaborative environments and multi-player games.

The avatar experience is an emersion experience. Deep learning changes your view of yourself. Desired change is towards being a more self-fulfilled individual. Sometimes this happens in a new place.

The evolution of the second school will create seeds for a future America where all individuals understand core and common foundations to science and higher mathematics. This understanding must occur in a context that acknowledges you as an individual.

Many current educational pathways lead to less and less capacity to be self-directed in primary learning experiences. It is proper that change agents use social media and three dimensional simulation worlds as an online platform.

Here the identity of self may be experimentally altered, so that peers see one's avatar as someone who is able to communicate about the foundational concepts in higher mathematics. Is that you?

What we gain

The system struggles with maintaining mass education, when what we could have is universal education at a far higher level than is currently being obtained.

This gain has national scope at a lower total cost. Tax revenues increase as incomes go up, but not federal expenditures. The second school is built on a business model that is self-sustaining. The model has not-for-profit characteristics.

What we perceive is a product of past experiences and expectations. There are, of course, individual differences, and these differences must find a means to cope within cultural contexts.

The effort required when one is not conforming is often overpowering, so accommodations are sought and found. One joins a group and through this group finds a sense of location, purpose, and meaning.

Or one isolates oneself from experience with an attempt to minimize the effort needed to be comfortable.

How to begin?

Attend and actively participate in class. Independence does not mean that you reduce your attendance.

Through social learning media, class participants at all levels work together to communicate basic content of specific coursework. The task we undertake together is to shift our mindset from one that is perhaps confused about what higher mathematics is to a mindset that understands deep learning methods.

This means that whichever class you are enrolled in and whomever is assigned to teach that class, you are asked to allow that situation and this book to work together to open these doors.

We ask you to see the task of learning to learn mathematics as more important than merely passing tests. Forgive the instructor when he or she is not meeting your needs. And forgive yourself.

Shared goals

As you begin to show additional promise on tests and in class your instructor's capacity to open doors will increase. However, initial expectations between the professor and the student are shaped in ways that must be accepted as being "as it is."

Deep learning theory develops a framework designed to shift the mindset involved, both your own and the system at-large. So your use of the methods could make your instructor take notice of improvements.

Our image of self is why people all around the world see Americans as non-math learners. You may consider specific changes in how you regard your study of mathematics.

In this way, you make a contribution to everything, everywhere.

Everyone's shared goal is to get a good grade and to rid yourself of social disdain felt about math class.

Five Topics

For students in advanced courses, we offer parallel studies using trajectories. An online mentoring community responds in a timely fashion.

You are the agent that drives events. You tweet and the system tweets back. If you do not learn to write math class notes, you will not be able to produce a trajectory.

In introductory studies, topics start in an interesting way. The first five are:

a) Re-learning arithmetic
b) Re-learning set theory
c) Learning for the first time about the set of real numbers

We then take the next step by considering:

d) Two-valued functions, and
e) Definition of a probability space.

Some Hidden Topics will be identified, and when identified may become part of a mentor service contract, extending you outside the boundaries of a manual.

Illustration of the Four-Step Method

The Four Step Method requires that you make up, or find an example or illustration of a focus topic. The steps need not be in the order presented below.

For example:

Step 1 (Illustrate): Consider the algebraic expression,

$$2\,(\,3x - 1\,)$$

This expression "simplifies" using the Laws of Arithmetic to the expression:

$$6x - 2$$

When we take any element from the expression's replacement set, this element will close the original algebraic form of the expression to produce an arithmetic expression.

The replacement set for expressions in one variable is the set of real numbers, not the integers, not fractions only, but the full remarkable set of all real numbers.

The same replacement element will close the simplified expression. If no error is made, then the two arithmetic expressions will be the same number.

The Four-Step Method (cont.)

Let 1/3 be selected to replace the variable *x*. Then the original expression becomes

$$2 (3 * (1/3) - 1)$$

The simplified expression becomes

$$6 * (1/3) - 2$$

The two arithmetic expressions are both the same number: i.e., 0.

Step 2 (Name): The focus topic is "The Replacement Set"

Step 3: (Revise; e.g., illustrate again): Consider the algebraic expression,

$$(2x - 1/3) * (3x - 1)$$

This expression "simplifies" using the Laws of Arithmetic to the expression:

$$6x^2 -3x + 1/3$$

If we use x = 1 as the replacement element then the two algebraic expressions are both closed to produce the number 10/9.

The Four-Step Method (cont.)

Remarkable as it may seem, fact is that the two core concepts of replacement set and solution set are largely absent in almost all non-STEM major students' minds.

It is as if they were never mentioned in previous courses.

Continuing with our narrative, we now produce an exercise in a number base other than base ten.

Because of how this is done, the illustration's statement exposes relationships between rational and irrational numbers.

Consider now:

((2.4) in base 10 is (?) in base 7?)

Upload your handwritten narrative to learning media. This might be the first time you communicate with any of the mentors, so try this and see what happens.

Four-Step Method (cont.)

Step 4: (Extend, i.e., talk about)

Develop in your own words a narrative about the nature of the replacement set. Use notation that is being shown to you by second school mentors or in face-to-face workshops.

Take what would seem to be a very easy topic and also a very deep topic and do what you can to expose the relationship between rational and irrational.

Does the mere fact of changing how a real number is represented change whether or not the number is rational or irrational?

What about the nature of being a prime integer? If we are referring to a known prime number, does changing how the number is represented change the prime property?

Can we represent 1/3 in a finite number of steps, simply by changing the base? Try base 6.

The Loop Exercise

Early in the illustration of the Four Step Method we saw an example of a Loop Exercise.

The narrative has a "perceptual" loop because, for example, an answer is obtained in two different ways: e.g., using 0 and then 10/9.

If you do not have strong arithmetic skills this weakness will show up when attempting to produce a loop exercise narrative about number base conversions. It is therefore essential that your arithmetic skills be strengthened.

If arithmetic in non-base 10 representations seems simple to you, then your skills are strong. We move on in search for that boundary between what you are comfortable with and what is uncomfortable to you.

Loop exercises free you from some external authority. One finds a *facilitating other* in your own actions.

Take responsibility for finding what is true about your own perception about mathematics. As you do this, it is possible to move just a little further to debunk any notion that freshman mathematics is not learnable or unimportant.

Topic Maps

Topic maps, concept maps, cognitive graphs, mind maps, and the nodal forest of topics – oh my, what has the world come to?

It is just a set of useful tools. We use lists of topics to find location. Trajectories help second school mentors and algorithms to make recommendations about what to study next.

Your true location is about you as an individual. And locations change when deep learning occurs.

Of course there are other factors involved in making the change you seek.

In order to learn hidden topics an individual generally needs,

1) Social reinforcement

2) A change in motivation

3) At least eight hours of short duration study.

Think about working self-made exercises for between 3 and 15 minutes each time. Over a few days, these short durations will add up.

Next topics

As you begin to master non-base 10 arithmetic exercises, we could introduce polynomial and rational functions. The learning process now becomes relevant to your college course:

a) A review of what a function is, both discrete and continuum domains and ranges

b) Composition is seen again, along with the notion of computational inverses with simple discrete functions

c) Restrictions of domain and range

d) Asymptotes

e) The introduction and use of linear functions to build rational functions

f) Finding the computational inverse of a linear function divided by a linear function (along with notion and theory about rational functions).

These topics are treated only after the individual has mastered number base arithmetic on the set of real numbers.

The journey begins

You have begun a journey. Your location is expressed from expert observations about your handwritten descriptions of topics.

Digital images of these descriptions are uploaded into the social learning media. We list topics, and separate that list into two. You develop detailed descriptions about that topic you find most challenging and a topic that you are uncomfortable about.

You maintain a private blog about how you feel.

Digital images are then placed into a database where artificial intelligence algorithms produce representations of individual mood, learning styles, resentments and other personal expressions.

A mentor will read your trajectory. The response he or she makes will be encoded as part of our continuing service to your learning process.

When you complete your work with us, all of this data is removed, even from the cloud. We do not need it, and neither will you.

Your particular journey

The objective of the second school is to provide digital resources that you may use at a time of your choosing.

An inquiry into any question could be how you start your experience with the second school. Or you could choose to use the methods suggested here to create a specific "scope" related strongly to that on which your school is testing you.

Scope is like a phrase entered into a search engine. But with trajectory scope you identify a part of the curriculum that is of interest to you.

"First Scope" is associated with the area of curriculum that you would like to start with.

Let us get started!

Part Three: Founding the Second School

Founding the Second School

Might social learning media make a permanent change in how America educates our children about mathematics and science? It is a well-known concern, with very few defined pathways towards resolution.

A number of embedded issues are involved: how to measure learning, how to provide interventions when behavior is not positive learning behavior, and others. Adaptive assessment is going to be a part of any resolution. But adaptation means change.

We cannot continue to simply lay down a linear program of study and ignore what is actually happening.

The consequences have produced a social phenomenon.

The 60% - 40% Problem

The core problem is not with the upper 40% of academic performers. Our colleges and universities do an excellent job with that group.

All too often academic careers end because of repeated failure in mathematics class. A good percentage of these failures can be traced back to individual experiences in school math classes.

Of course, the issue is larger than this.

An analysis was made about the nature and structure of an American enterprise that expends almost one trillion dollars in local, state and federal taxes per year.

This analysis explains why an independent American Education Bridge should be funded at the federal level.[17]

Short versions, often one to five pages in length, focused on three pillars that might be constructed. Through this construction we might hold together a single, but independent, system of instruction. A national education bridge between high school and college might be deployed.

---

[17] Prueitt, Paul Stephen (2015) *Learning Media, the American Education Bridge between High School and College. Applied Knowledge Sciences Publishers*

Three Pillars

The three pillars are: 1) technology, 2) pedagogy, and 3) behavioral neuroscience. Discussions about a common architecture[18] [19] were presented in various academic forums.[20] [21] [22]

---

[18] Prueitt. Paul Stephen (November 14 2013) Combining Deep Learning Method and Self-Directed Inquiry, Foundational Concepts and Challenges. NSF sponsored talk at Texas Southern University.

[19] Prueitt, Paul Stephen (2012) New Proposal for Educational Reform, accepted Interdisciplinary Research, Education, and Communication (IDREC 2012) March 25-28, Orlando Florida as part of The 3rd International Multi-Conference on Complexity, Informatics and Cybernetics: IMCIC 2012

[20] Prueitt, Paul Stephen (2013) Deep Learning Methods and Adaptive Assessment, presented at University System of Georgia Teaching and Learning Conference: Best Practices for Promoting Engaged Student Learning

[21] Paul Stephen Prueitt (April 17th 2015) A New Data Source for Design-Based Research on Deep Learning Strategies, presentation to the Mathematics Education Seminar at Texas State University

Paul Stephen Prueitt (Feb. 6th 2015) Deep Structure Architecture for Machines and Humans, presentation to the Mathematics Education Seminar at Texas State University

[22] Paul Stephen Prueitt (February 2015) Deep Architecture for Human and machine Learning, seminar presented at Computer Science Seminar at Texas State University

Presentations over recent years have communicated how Prueitt's early insights on educational reform evolved.[23] [24]

The central feature he refers to as "stratified theory" was realized as an architectural design for distributed computing systems.[25] [26] [27] [28] [29] [30] [31]

---

[23] Prueitt, Paul Stephen (1988) Discrete Formalisms and the Ensemble Modeling of the Instructional Process, published in the proceedings of The Fifth International Conference on Technology and Education, Edinburgh, Scotland, March 1988.

[24] Dissertation: Prueitt, Paul Stephen (1988) Some techniques in mathematical modeling of complex biological systems exhibiting learning, PHD Thesis, in Pure and Applied Mathematics, University of Texas at Arlington Press

[25] Prueitt, Paul Stephen in Service Technology Magazine – (Jan 2015) New Data Analytics

[26] Prueitt, Paul Stephen in Service Technology Magazine – (July 2013) Service Oriented Architecture and Data Mining: a step towards a Cognitive Media?

[27] Prueitt, Paul (2011) - "Digital Instrumentation and the Measurement of Experience" SOA Magazine

[28] Prueitt, Paul Stephen (2011) Systems Science and Service Computing, Published Dec 14th 2011, Service Technology Magazine.

[29] Prueitt, Paul (2009) - "The Service Engine: Structured Communication using Modern Service Technologies" SOA Magazine

Stratified theory is different from classical theories of natural science. [32] A core underlying assumption is that natural systems self-organize into "isolated" strata.

Organizational strata are observed in nature. [33] For example, each person's speech is produced by a small set (less than 50) of "motor reflex arcs."

Our human-in-the-loop plus AI work processes create behavioral and knowledge profiles.

---

[30] Prueitt, Paul Stephen (2009) Articulating SOA in the cloud, SOA Magazine

[31] Stratified theory and service architecture were realized in the work of Sandy Klausner, founder of CoreTalk. That work may be reviewed at coretalk.net. Starting in 1998, Prueitt worked on several parts of the CoreTalk architecture and theory.

[32] Prueitt, Paul S. (1995a) A Theory of Process Compartments in Biological and Ecological Systems. In the Proceedings of IEEE Workshop on Architectures for Semiotic Modeling and Situation Analysis in Large Complex Systems; August 27-29, Monterey, Ca, USA; Organizers: J. Albus, A. Meystel, D. Pospelov, T. Reader

[33] Prueitt, Paul S. (1996b). Is Computation Something New? published in the Proceedings of NIST Conference on Intelligent Systems: A Semiotic Perspective. Session: Memory, Complexity and Control in Biological and Artificial Systems. IEEE October 20-23.

Profiles are abstractions produced by separating what we see as commonalities across multiple instances and the composition of these commonalities into individual, but abstract, representations of you .

Pedagogy defined and tested

Prueitt's effort at Texas State was sufficient to test instructional pedagogy in non-STEM mathematics courses. Over 600 students were enrolled in classes he taught from August 2014 – May 2016.

A summary of a survey is presented below. Each question's response is given by selecting one of the following:

strongly disagree, disagree, no opinion, agree, strongly agree.

The summary shows the average response, developed by selecting *-2, -1, 0, 1, 2* as representations of *Likert scale* marks.

Here we must address an un-solved problem.

Our best universities cater to those who are in the top 40% of performers. By focusing on the 60%, the class is slowed. In very competitive environments this is not considered a good thing.

If you already have the proper view, then the methods may be a waste of your time. You are in college to be part of the top 40%.

Deep learning methods are designed to change one's view about mathematics. If one's views are already ok, there is nothing to be gained.

Polling Instrument

Please do not put your name on the poll, as this is anonymous.

Course 1319: TR at 9:30

Purpose of this Poll:

Dr. Prueitt has used a specific pedagogy whereby students are challenged to write about the mathematical topics in the course curriculum. The pedagogy might produce more long-term remembrance of how mathematical exercises are completed. Or the pedagogy might be inconsistent with the purpose of non-STEM mathematics courses.

This poll measures

1) Student satisfaction / dissatisfaction with experience in Dr. Prueitt's classes
2) Student understanding of the nature and purpose of the pedagogy
3) Student estimation of success of the course
4) Student comparison to other non-STEM mathematics courses
5) Student sense of comfort with math class.

Polling Instrument: April 2016

Overview: Your view about the nature and quality of courses offered by Texas State is important to us. You took a course instructed by Dr. Prueitt.

We are interested in your opinions about that course. Please indicate your response to a series of statements in two ways.

First by marking on the line:

Strongly Disagree Dis-agree No Opinion Agree Strongly Agree

____

in response to each of twelve statements.

Then, please write an extended comment about your viewpoint. Please take your time.

(1) The methods used by Dr. Prueitt were different from what I expected.

Strongly Disagree Dis-agree No Opinion Agree Strongly Agree

____________________________________1.2 ____________

(2) These methods were at first difficult to understand.

Strongly Disagree Dis-agree No Opinion Agree Strongly Agree

_____________________________0.4____________________

(3) Eventually I did understand what was expected in this class.

Strongly Disagree Dis-agree No Opinion Agree Strongly Agree

____________________________________1.2 ____________

(4) Once understood the methods seemed not necessary for me, and a waste of my time.

Strongly Disagree Dis-agree No Opinion Agree Strongly Agree

__________________-0.6_______________________________

(5) My understanding of the curriculum in this course was improved more than I expected, and because of the pedagogy.

Strongly Disagree Dis-agree No Opinion Agree Strongly Agree

_______________________________________1.4 __________

(6) I believe that my previous experiences in mathematics class have been mostly positive, and that I was adequately prepared to enroll in this class.

Strongly Disagree Dis-agree No Opinion Agree Strongly Agree

_________________________0.06 ______________________

(7) I believe that the methods used in Dr. Prueitt's class could be improved if social media was developed to support my 7-24 communication with the class and with the instructor.

Strongly Disagree Dis-agree No Opinion Agree Strongly Agree

_________________________________ 1.08 _____________

(8) I have a clear understanding of what Dr. Prueitt refers to as 'Deep Learning Methods.'

Strongly Disagree Dis-agree No Opinion Agree Strongly Agree

____________________________________1.37 ___________

(9) I have a clear understanding of how my sense of location within my study of a curriculum might be improved using topic trajectories.

Strongly Disagree Dis-agree No Opinion Agree Strongly Agree

__________________________________ 1.34____________

(10) I do ***NOT*** believe that these methods should be used in this type of course.

Strongly Disagree Dis-agree No Opinion Agree Strongly Agree

__________________-0.6 _____________________________

11) I would like to see these methods developed as part of an effort to improve non-STEM major mathematics offering at Texas State University.

Strongly Disagree Dis-agree No Opinion Agree Strongly Agree

_________________________________ 1.25 __________

A twelfth question was not included, but may be included in an upcoming survey.

(12) I feel that mini-course, teaching deep learning methods, should be commercialized into adaptive mentoring products

Strongly Disagree Dis-agree No Opinion Agree Strongly Agree

____

–

Poll Summary

The April 2016 poll of one class section indicated successful implementation of new and un-expected pedagogy. The poll is consistent with departmentally administered student ratings of Prueitt's classes.

Student comments may be reviewed under agreements to not disclose details. However, in summary these comments provide evidence that widespread support for deep learning methods arises from students after an initial period of time.

It has also been true that first reactions are often negative due to student expectations being both well-established and strongly held.

And high achievers sometimes complain. These complaints are a source of difficulty for innovators.

In a sense, this behavior has served to limit the number of individuals who are "good at math".

Some complications

Deep learning methods do not need to be learned if one already has an active learning process. But if someone thinks deeply about mathematics he or she might become a leader in the class, and possibility a certified second school mentor.

All we have to do is begin the development of a dedicated social learning media following well-established patterns in code design. We might even extend the use of the Knewton® platform.[34]

We have proposed that a national independent social learning media be dedicated to education. As a core function, it would allow student trajectories to be messaged, much like on Twitter® or other popular social media feed.

Analytics on many trajectories reveal where student populations are in location. One can then orient algorithms and workflow to refine an understanding of what is occurring.

---

[34] Knewton® is perhaps the leading adaptive learning technology provider.

From properly designed social learning media one may acquire de-individualized data about exactly what is occurring as hundreds, or hundreds of thousands, or millions, of students communicate with each other about concepts being studied at any given moment in their math class.

The challenge is to engage the 60%,
while not boring the 40%.

Developing supporting technology

Central to understanding deep learning methods are *sense of location* and *action-perception cycles.*

Social learning media orients towards a class, or an individual student. It looks at handwritten expressions from students and recommends what types of exercises a student should look at next.

Of course there are always choices: should we recommend this or that?

Writing about mathematical topics is not something non-STEM majors are expecting. However, several very critical concerns relevant to the learning task are addressed in this way.

There is internalization of material as a student practices with trajectories. Because avatars are anonymous, the student is able to communicate with his or her sense of self.

Summer and autumn 2016

In spring 2016, technology was developed to integrate with the university course management system and e-mail at Texas State University.

The next step involves designing and integrating a system like the Knewton® platform. In this platform, artificial intelligence processes incoming data and references ontological models about target scope. The inference engines produce recommendations.

These programs "learn" by using algorithms to separate categories. Location is placed as being within a specific category or group of categories.

Far simpler software supports our summer and fall pilots. But there will be no AI, just simple support for students sending in pdf files and insightful mentors responding using email.

This opens up employment opportunities for advanced undergraduates as certified second school mentors.

Our mentors are well-trained in the methods and very knowledgeable about the freshman curriculum. We have regularly-scheduled study halls in various locations such as in a coffee or teahouse, and at specific spots around the campus.

Sample illustrations

Foundations Assignment #2

Part #1

List of Topics: - Composition of Functions
- Compositional inverses
- Exponential functions
- log functions

Comfortable
- Composition of functions
- Compositional inverses
- Exponential functions

not comfortable
- log functions

↓

I feel as if I havent learned enough of the steps for those problems to be able to solve one completely on my own.

The first step in building a trajectory is making a list of topics that you feel represent the curriculum set by the college or university. You then separate that list into two parts: that which you are comfortable with and those topics that you are not.

part #2 = Detailed description of most complicated "comfortable" problem/topic

one of the hardest topics that I am comfortable with is compositional inverses

example: $f(x) = x^2 - x + 3$ & $g(x) = 2x - 1$

a. $(f \circ g)(x)$
b. $(g \circ f)(x)$

a. $(f \circ g)(x) = f(g(x))$
$= f(2x-1)$
$= (2x-1)^2 - (2x-1) + 3$
$= 4x^2 - 4x + 1 - 2x + 1 + 3$
$= 4x^2 - 6x + 5$

b. $(g \circ f)(x) = g(f(x))$
$= g(x^2 - x + 3)$
$= 2(x^2 - x + 3) - 1$
$= 2x^2 - 2x + 6 - 1$
$= 2x^2 - 2x + 5$

$(f \circ g)(x) = 4x^2 - 6x + 5$
$(g \circ f)(x) = 2x^2 - 2x + 5$

This is followed by a detailed description of the most challenging of all of the topics listed by you in your topic enumeration.

Avatar profiles

Avatar profiles represent learning-related characteristics within a subpopulation of students who are sharing the same or similar feelings about their math class.

I think where I went wrong w/ log functions
is that I did not comphrehend them well from the
start. I just started looking @ an introduction to
log functions and it is starting to make some sense.
I start to get overwhelmed during tests because
I get confused and don't know what to do. but I
am trying to fix that by working extra on learning.

Social learning media has the ability to encourage individuals to interact through an *abstraction layer*.

A mentor avatar is created which interacts through adjustments made by both the AI programs and human mentors working with that sub-population of students.

Narratives

A narrative blog-type entry allows the individual to define what his or her concerns are. Not only are personality types correlated with handwriting, but also the handwriting may be automatically re-expressed as ASCII text.

Part #4

1) "Where am I in the learning process?"
Most of these topics I am comfortable with, which is good, but I can always use extra practice and go over these topics more regularly.

2) "What are my feelings about this class?"
I was confident in this class prior to my test, but after this past test, it made me realize I may need to leave out more time of studying to study for this class

3) "What is my understanding of the learning strategy we are using?"
When I was first introduced to the narratives strategy last semester, I did not know what to think of it at first. Now, I like the narratives strategy cause it allows me to break down each topic more in depth and understand what I am learning, in a different way.

Text of this type may be processed by very mature linguistic analytics.

Movement in Topic Spaces

Once a sense of location is found, deep learning begins. Even then we must move slowly. Moving slowly is important due to predictable consequences from past learning experiences.

For example, experiences learning math are often incomplete.

For whatever reasons, a few essential topics are often hidden from view. That is, if some students are struggling with the math class.

This essential point is understood by those who are not doing well in math class.

Behavioral Analysis

Relevant analysis about behavior must include both systemic reactions by typical departments of mathematics and by various types of individual students.

The current system of mathematics education does an excellent job of producing mathematicians, but at the expense of specific un-intended consequences. One such consequence is an almost universal disdain towards math class, particularly by American students.

Math disdain is simply due to our not yet having evolved pedagogy to a point where it is designed to encourage individuals faced with the possibility of failing a course. This has been where some of the disagreement has developed between founders of the second school and mainstream mathematicians.

The system does its thing and those whose individual constitutions agree are selected as being math-capable. This does not mean, however, that this selection is fair or optimal.

Applying Deep Structure to how we educate

The Manual has several descriptions about key elements in deep learning methods as Prueitt defined them in classroom practice at Texas State University. We would like to give a slightly more formal definition.

Deep learning is assumed to occur only when an underlying *organizational substructure* to human awareness is re-structured.

Let us suppose that most, perhaps over 80%, of all non-STEM majors have deep structural problems caused by experiences in school or previous college math class. By this, we mean all college students who are not STEM majors.

Deep learning methods are used in overcoming specific problems.[35] [36] There is self-discovery as well as change in sense of location. One discovers characteristic constitutional elements composed to produce one's behavior.

---

[35] We use stratification theory as a model of any complex; e.g., natural, system.

[36] Rosen, Robert (2012), "Anticipatory Systems; Philosophical, Mathematical, and Methodological Foundations, 2nd Edition, Springe

Support from Natural Science

If our assumptions are in line with reality, there will be behavioral correlates. These might be enumerated and discussed in scholarly terms. This was communicated in two seminars given by Prueitt in 2015. The issue is measurement of learning.

Theoretical constructs derived from natural intelligence literatures [37] suggest that an action-perception cycle is involved in human awareness.

The human mind develops an ability to judge reoccurring situations. A good professor will make judgments about the class as a whole or about an individual student. Of course if judgments are to be acted on, one needs to be able to change schedules accordingly.

Adapting in this way is generally frowned upon. Students want, for example, to know when tests are to be administered.

---

[37] In particular the work of James Gibson:

Gibson, J.J. (1950). The Perception of the Visual World. Boston: Houghton Mifflin.

Gibson, J.J. (1960). The Concept of the Stimulus in Psychology. The American Psychologist 15/1960, 694–703.

Gibson, J.J. (1966). The Senses Considered as Perceptual Systems. Boston: Houghton Mifflin. ISBN 0-313-23961-4

Fast thinking and slow thinking

Stratified theory suggests that perception and action may operate on a fast or slow time scale. The fast time scale can be separated from the slow time scale through what is called *blocking*.[38]

Blocking occurs when students cram for tests. Separating is seen as inducing deep learning by slowing thought so that one's learning about underlying concepts are accommodated by the individual's mind.

Part of the method involves learning how to write, onto unlined paper, about mathematical topics. These handwritten detailed descriptions are then scanned to a pdf file and uploaded to a social learning media platform.

Response should be fast, but processed slowly. This type of assessment/recommendation engine is not yet available unless a human is in the loop.

Again, part of the function of the second school social learning media is to provide a means to reduce the time between when a student submits a trajectory and the media sends something back.

---

[38] Kahneman, D. (2011) *Thinking, Fast and Slow*, Farrar, Straus and Giroux, ISBN 978-0374275631

Topic Trajectories

Four pages folded longwise with name ONLY on the outside.

Each Trajectory should have a time stamp.

Page One is a list of topics, and then a separation of the list into the two parts. The list does not have to be comprehensive, just useful to you. The list has a *scope*, which is sometimes a set of homework exercises, or a textbook section or group of sections. Negotiating the scope is part of "finding location."

Page Two is a detailed description about the most challenging of all topics listed in the comfortable list. The description should be about things related to the topic and will use at least one exercise from the textbook. We use notation taught in class to help us express more than just the question and the answer. Choose the topic as a means to communicate what you know.

Page Three is a detailed description about one of the topics in the "not-yet-comfortable-with" list. The purpose here is to communicate what you know about the topic, and also to reveal what makes you uncomfortable. Where and how do you feel confusion?

The Descriptive Blog

Page Four is a handwritten message to the instructor, and to yourself, about

(1) Where "I" am in the learning process,

(2) What are "my" feelings about this class, and

(3) What is "my" understanding of the learning strategy we are using?

This is like a blog entry.

Listing topics is an exercise in which the individual makes a list based on a personal sense of how this list is useful to the individual. He or she will use this list to find the boundary between what is comfortable and what is not comfortable.

Scope and Enumerated Lists

So the topic list has scope. Once this scope is established, the list is separated, by the individual, into two parts: those topics that one is comfortable with and those topics that one is not yet comfortable with.

Repeating the listing exercise without looking at previous lists creates movement within one's sense of what one is studying. One is allowed to look into the book, talk to others, or not.

Detailed description of what one is comfortable with shows mastery over topics of one's own choosing. One's ability to use advanced notation is enhanced.

Analysis about individual learning barriers becomes easier. The description of what one understands about a topic, listed as uncomfortable, often creates movement in one's understanding space.

Boundaries

Each trajectory has a blog entry reporting one's feelings and the description of a boundary between what one is comfortable with or not. This communication is under the control of the individual and informs the teacher in a precise fashion.

Students are asked to create a series of these trajectories as a means to self-report on learning.

In response to the instructor reading the trajectory, a recommendation is made as to a "next assignment."

Creating Action-Perception Cycles

Action-perception cycles reflect a physical reality involved with human awareness. Technology in social media has been attempting to produce a digital technology that would adaptively assess what the individual wants: his or her "location" in some type of representation space.

A proper assessment creates a digital profile that is then useful in recommending to the individual something to buy, or to otherwise consume.

Similarly, deep learning methods produce an internalization of elementary notions in a mathematics curriculum. This is occurring for those who are being successful, and not occurring for those who are not being successful in class.

Our desire is have a means whereby the individual could communicate where he or she is "located" in a representation space of the topics in the curriculum. This starts a process that, if allowed to continue, increases the chance for success for that individual in math class.

Examples of handwritten descriptions

The following screen shots are from one topic trajectory series. Each image is distorted so as to disguise the identity of the individual.

Quadratic Functions
Quadratic Equations
Quadratic Inequalities
Quadratic Formula
Vertex form
Graphing a quadratic function

3/24/16

Comfortable

Quadratic Functions
Quadratic Equations
Quadratic Inequalities

Not Comfortable

Quadratic Formula
Vertex form
Graphing a quadratic function

Handwritten description examples (cont.)

Quadratic Equations

If a, b and c are all real numbers with $a \neq 0$, then the function $f(x) = ax^2+bx+c$ is a quadratic function and its graph is a parabola.

So, if an equation is in Standard form, $f(x)=ax^2+b+c$, all the numbers are all real numbers and it is a Quadratic function.

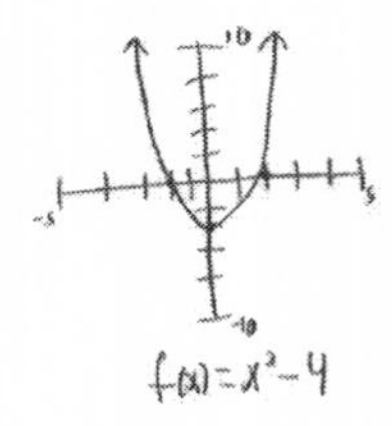

A student creates a list of topics, and then separates this list into two parts. Then, following instructions, this student wrote about what he or she is comfortable with.

Handwritten description examples (cont.)

Vertical and horizontal shifts

$y = |x|$

$y = |x| + 4$

$y = |x| - 5$

So for this example we are using the basic elementary function of Absolute value. On a vertical shift the x intercept does not shift as the y intercept is being moved. The horizontal shift is just the opposite as the y axis remains same and the x axis is shifted.

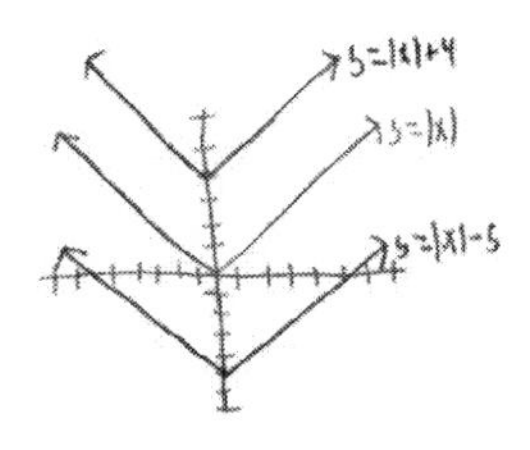

Translations of Graphs

The student is expressing comfort in the absolute value function and in translating the graph of a function up or down.

Handwritten description examples (cont.)

Basic Elementary Functions
Vertical shifts
Horizontal shifts
Reflections
Streches
Shrinks
Combining graphs
Piecewise functions

3/23/16
12:30pm

Comfortable
Basic Elementary Functions
Vertical shifts
Horizontal shifts

Not comfortable
Reflections, Streches, Shrinks
Combining graphs
Piecewise functions

A student then selects one of the topics he or she is not comfortable with and writes about this topic and concerns.

Handwritten description examples (cont.)

As I continue in this class the concepts have gotten harder
which with my previous math classes it adds stress. For me personally
it is eass to feel overwhelmed by math. As I continue to
complete these topic trajectories it is making me feel less
and less overwhelmed. I personally have taken a liking to this
learning process and teaching style. My feeling about this
class are great I understand more about math then I ever have.
My understanding with this learning strategy is meant to
take the stigma out of math and allow me to study
for math like how I study for any other class.

In many instances, the intent is not yet seen. The student is writing about his or her sense of self.

SO, that was straight from the
book it would help to write it out.
I am still a little confused.
I remember this from previous
math classes but I never fully
understood.

Here the student expresses how he/she feels about math and about the class.

Handwritten description examples (cont.)

Quadratic formula

An x intercept of a function is also called a zero of the function. The x intercept of a linear function can be found by solving the linear equation $y=mx+b=0$ for x, $m \neq 0$. Similarly, the x intercepts of a quadratic function can be found by solving the quadratic equation $y=ax^2+b+c=0$ for $x, a \neq 0$. Several methods for solving quadratic equations. The most popular is the quadratic formula.

$$x = \frac{-b \pm \sqrt{b^2-4ac}}{2a}$$

$$-x^2+5x+3=0$$

$$x = \frac{-(5) \pm \sqrt{5^2-4(-1)(3)}}{2(-1)}$$

$$= \frac{-5 \pm \sqrt{37}}{-2}$$

$$= -0.5414 \text{ or } 0.5414$$

Imagine this communication as part of a discussion between an individual enrolled in a college math class and someone who has deep knowledge about not only the subject matter, but also why people have difficulty with the quadratic formula.

Handwritten description examples (cont.)

Quadratic Equations

If a, b and c are all real numbers with $a \neq 0$, then the function $f(x) = ax^2+bx+c$ is a quadratic function and its graph is a parabola.

So, if an equation is in standard form, $f(x) = ax^2+b+c$, all the numbers are all real numbers and it is a Quadratic function.

10

-5 5

-10

$f(x) = x^2 - 4$

The next part of his/her detailed description appears to indicate comfort with the notion of a quadratic equation and the corresponding graph.

Handwritten description examples (cont.)

So I belive at this point in the semester I have learned more
about then in the past, more importantly I belive that what I am
learning is sticking and that I will be able to recall the
material better in the future. I think right now when it comes
to quadratic functions I am not 100% clear on this
topic, I can sit and figure out a problem with the help of the
book but I couldnt tell you how I got that answer. So
I think this class is really helping me remove the
stigma I have had about math. I like that I am
begging to transition that I can study math like history
and I honestly belive that it creates a better understanding
Personally. I just need to put more time in. I have
been working too many hours recently and I have been
falling behind. My work finally hired new people so my
hours will thankfully decrease, and I should be able to
dedicate more time to this class and the others.
I really enjoy this class and will be working harder
and completly remains the stigma of math class

One sees utter complexity in an individual's internal thinking about the math class. How is complexity of this type to be measured and interpreted?

Handwritten description examples (cont.)

However

$y^2 - x^2 = 9$

This is where my confusion
begins. I dont know why I
get confused when exponents are
involved but they throw me off.
I know the answer is that it is
not a function becuaes the
equation has two different outputs,
I just struggle on fully understanding
How to reach the answer.

The student's sense of confusion is now apparent. Discovering how to properly guide the student with this level of increased depth of measurement is a profound challenge.

Our assessment/recommender engine must interpret several contradictions. Does the student feel comfortable about any of these concepts, and if so, which ones? This type of profile construction is beyond current best practices in college.

Handwritten description examples (cont.)

functions and equations

$$4y - 3x = 8$$
$$+3x \qquad +3x$$

$$\frac{4y}{4} = \frac{8}{4} + \frac{3x}{4}$$

$$y = 2 + \frac{3}{4}x$$

This I understand, how the equation is solved to equal one outcome and is a function.

"Ah, I see now!"

Of course, the question is how long this understanding will be accessible by the individual.

Handwritten description examples (cont.)

Point by Point plotting

$y = 15 - x^2$

x	-3	-2	-1	0	1	2	3
y	-6	-11	-14	15	14	11	6

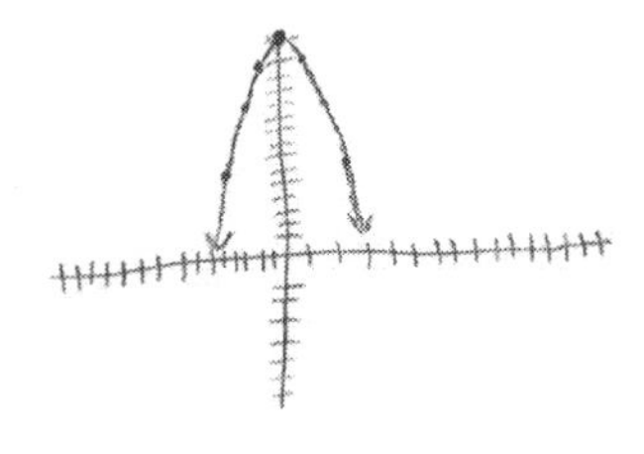

So with this topic the first thing is that it is appent I need to work on my graph drawing skills for point by point plotting it is taking an equation that is more difficult to graph than Ax+By=C. To graph y=15-x² it is a good place to start with a table by Substituting x with a number and then solve the equation. By doing this it allows me to easily plot and sketch the graph

In the screen-captured images above we see self-directed exploration of the topics within the student's self-defined scope. The individual is completing a trajectory. Below is a second trajectory, one created the day before.

Handwritten description examples (cont.)

for the most part I am fairly comfortable with this
Chapter. which is good because I know that the rest
Of the chapter is going to build on these basic principles.
espacilly when it comes to the Basic elementary functions.
I am glad I went back and re learned them becuase it
was causing alot of confusion for me. leading up to
this test I am really focusing on mastering chapter
2 because I belive that will help me succed in 1329.
The more I become with this deep learning the more
I have become to truly understand this matirual.

Motivational talk is seen, and concern about an upcoming exam.

Index

ABOUT THE AUTHOR

Professor Paul Stephen Prueitt has taught mathematics, economics, physics and computer science courses in community colleges, universities, and four-year colleges. He has served as Research Professor in Physics at Georgetown University and Research Professor of Computer Science at George Washington University.

He served as Associate Professor or Assistant Professor of Mathematics at HBCUs in Virginia, Tennessee, Alabama and Georgia.

Prueitt was co-director of an international research center at Georgetown University (1991-1994). He is a NSF reviewer and Principle Investigator. He served for over a decade as an independent consultant focused on information infrastructure, software platforms and intelligence algorithms. He consulted on national intelligence software platforms and continues this work under private contracts.

His post-Master's training in pure and applied mathematics focused on real analysis, topology and numerical analysis. His PhD was earned in 1988 from The University of Texas at Arlington.

The dissertation was developed using differential and difference equations as models of neural and immunological function. He has over sixty-five publications in journals, books or as conference papers.

Motivated by a desire to understand the nature of the American educational crisis, he served for seven years at Historical Black Colleges and Universities, including an open door minority serving institution in Atlanta Georgia.

Made in the USA
Columbia, SC
10 August 2021

43042523R00100